Great Ideas of Science

ATOMIC STRUCTURE

by Rebecca L. Johnson

Twenty-First Century Books
Minneapolis

With special thanks to Carol Hinz, whose consummate skill as an editor is evident on every page

Twenty-First Century Books
A division of Lerner Publishing Group, Inc.
241 First Avenue North
Minneapolis, MN 55401 U.S.A.

Website address: www.lernerbooks.com

Library of Congress Cataloging-in-Publication Data

Johnson, Rebecca L.
    Atomic structure / by Rebecca L. Johnson.
       p.   cm. — (Great ideas of science)
    Includes bibliographical references and index.
    ISBN-13: 978-0-8225-6602-1 (lib. bdg. : alk. paper)
    1. Atomic structure. 2. Matter—Constitution. I. Title.
QC173.4.A87J64 2008
539.7—dc22   2006018992

Manufactured in the United States of America
1 2 3 4 5 6 – DP – 13 12 11 10 09 08

TABLE OF CONTENTS

DIVIDING MATTER

Imagine you have a bar of chocolate—and lots of friends. One of your friends asks if you'd split your chocolate with him. You break the bar in two and give him one half. Before you can take a bite from the remaining half, however, another friend comes along. She asks if she can share your chocolate. So you divide what's left and give one-half to her. Now imagine that this process keeps going. . . .

You continue dividing your remaining piece of chocolate and giving one of the halves away. After several more divisions, you'll be left with a very small piece of chocolate—so small it wouldn't make much sense to share.

But could you keep on dividing it? Let's say you ground up that remaining piece of chocolate into a small pile of chocolate powder. You could divide it into two smaller piles. Then you could halve one of the smaller piles and keep repeating the dividing process.

Eventually, you'd be left with just one speck of chocolate. But if you had a microscope and some precise tools, you could continue dividing that speck into smaller and smaller bits.

Could this division process go on forever? Or would you eventually get down to some fragment of matter, some kind of fundamental particle, that couldn't be divided any further and still be chocolate?

Puzzling over such things may not keep you awake at

More than 2,500 years ago, philosophers speculated about what happens to gold when it is divided again and again.

night. But this question was an important one for Greek philosophers more than 2,500 years ago. What happens to a substance such as gold, they wondered, when a chunk of it is cut in half over and over again, into smaller and smaller pieces? Is there a limit to how small a piece of gold you can have?

In trying to answer these questions, the philosophers began a search for truth that eventually led to the discovery of atoms—the building blocks of all matter. And atomic research revealed the subatomic particles that make up atoms themselves. Our current understanding of atomic structure is built on the insights, ideas, and experiments of hundreds of scientists. Some were chemists. Many were physicists. Others were mathematicians. Their investigations into atoms and atomic structure spanned many centuries. And the search isn't over. Research into subatomic particles and their nature continues on many fronts in laboratories around the world.

The story of how we came to know about atoms and their structure is complex. But it began with people wondering about the world around them. What are the fundamental ingredients that make up all the matter in the universe, from plants and animals to oceans and stars? In other words, what is the nature of matter? Of what is it formed? And how do those fundamental bits of matter work together to make our universe what it is?

CHAPTER 1

EARLY IDEAS ABOUT ATOMS

The first person on record to ask what matter is made of—and to suggest an answer—was Thales of Miletus. He lived in ancient Greece from 624 to 546 B.C. Thales proposed that the fundamental substance out of which all matter was made was water. Water, the philosopher believed, had the power to change into anything in the universe.

During the next century, other Greek philosophers proposed that air or fire might be the fundamental substance. Empedocles suggested everything was formed from water, air, fire, and earth. These four substances came to be thought of as the four basic elements and the source of all matter.

Two other philosophers—Leucippus and his student Democritus—asked whether a substance could be endlessly divided. Leucippus concluded that at some point, indivisible fragments would be reached. The fragments could not be broken down anymore and still retain the characteristics of the original substance.

Leucippus called these fragments *atomos*, meaning "indivisible."

Democritus expanded on his teacher's atomic hypothesis. He stated that atoms were also solid and indestructible. Furthermore, he believed that the universe contained only atoms and empty space, which he called the void. Democritus also argued that atoms differed in shape and size. As atoms clustered together in different configurations to create matter, these slight variations yielded different kinds of matter with different properties.

The philosophy that all matter was formed from atoms came to be called atomism. However, the idea was never widely accepted in the ancient world. Aristotle, one of the most influential of all Greek philosophers, rejected atomism. He believed that only things that could be observed could be shown to be true. And since atoms could not be seen, their existence could not be proven by observation.

ARISTOTLE AND ETHER   To the list of water, air, fire, and earth, Aristotle added one more fundamental substance, ether. Aristotle believed ether, which in Greek means "blazing," made up the Sun, the stars, and other bright objects in the night sky. The philosopher also thought ether filled the apparent emptiness of space. With Aristotle's addition of ether, the list of basic elements in the ancient world grew from four to five. Belief in these elements endured well into the Middle Ages (A.D. 500–1450), although the earthly elements received the most attention.

Largely because of Aristotle's influence, atomism gradually fell out of favor. One of its rare supporters was the philosopher Epicurus. A follower of his, Lucretius, wrote *On the Nature of Things*, a poem that restated the basic concepts of atomism. But after his death around 50 BC, Lucretius's writings were all but forgotten.

In the centuries that followed, just one copy of Lucretius's poem survived. Following the invention of the printing press in the 1400s, however, that one copy was transformed into many. A new generation of thinkers read the poem and considered atomism with fresh eyes. While ancient Greek philosophers had tried to answer questions about the natural world using only logic and reasoning, scientists of the 1400s carried out experiments to test new ideas.

THE ALCHEMISTS

Atomism caught the interest of many scholars. They speculated that the four earthly elements (water, air, fire, and earth) could be formed from just a few different types of atoms. That would mean, then, that different arrangements of atoms would yield all the substances found on Earth. For example, particles that made up the element earth could combine in one way to form gold. Combined in other ways, the same particles might form mercury, carbon, or lead.

If that were true, of course, then lead could be turned into gold—if you knew the right way to rearrange the particles. This idea spawned alchemy, the manipulation of chemicals in the quest to change one substance into another. Alchemy flourished through the 1700s. Those who practiced it were known as alchemists.

Some alchemists were crackpots, magicians, and schemers who hoped to turn worthless substances into precious metals. But many others were thoughtful people who wanted to learn more about various substances and how they interacted. Alchemists played a major role in advancing understanding of basic chemistry. They discovered the properties of many chemical substances and showed how certain types of chemicals reacted when combined.

BOYLE'S EXPERIMENTS

Robert Boyle was an Irish alchemist who believed in atomism. Boyle performed carefully controlled experiments. He kept detailed notes so his experiments could be repeated precisely. Boyle proposed that the true elements were not water, air, fire, and earth. Instead, he suggested they were substances that could not be broken down by chemical manipulation into simpler substances, such as lead, iron, and sulfur. He explained his reasoning and results of experiments supporting this hypothesis in his book *The Sceptical Chymist*, published in 1661.

One set of experiments that Boyle carried out resulted in some of the first concrete evidence that atoms, although unseen, truly did exist. Boyle used a glass tube shaped like the letter *J*. It was closed at the short end but open at the tall end. By adding mercury to the tube, Boyle trapped a small amount of air in the short end. When he added more mercury (increasing the pressure exerted on the trapped air), he noticed that the volume of the

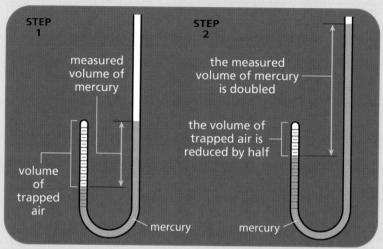

Boyle's J-Tube Experiment

STEP 1

measured volume of mercury

volume of trapped air

mercury

STEP 2

the measured volume of mercury is doubled

the volume of trapped air is reduced by half

mercury

Robert Boyle determined that adding liquid mercury to a glass tube would compress trapped air into a smaller space. Boyle and other scientists theorized that the air became compressed because the pressure forced the atoms making up the air closer together.

trapped air decreased. In other words, the air was compressed into a smaller space. Removing some of the mercury—and so relieving some of the pressure—caused the volume of trapped air to increase.

How was it possible for air to be compressed? Boyle realized that if air was made up of minute particles— atoms—that were separated by space, then perhaps those atoms were being forced closer together when pressure was applied, thus decreasing the volume of the air sample. Reducing the pressure would allow the atoms to spread farther apart, increasing the volume.

BOYLE'S LAW   After several trials with his J-tube experiment, Boyle noticed a mathematical relationship between the weight of the mercury column (the amount of pressure exerted on the trapped air) and the air's volume (the amount of space the air occupied). Double the pressure, and the volume of the air was halved. Halve the pressure, and the volume was doubled. This inverse relationship between pressure and volume is called Boyle's law.

Boyle's experiments on air gave powerful support to the idea that matter consisted of atoms surrounded by space.

FRENCH EXPERIMENTS

More evidence for atoms came in the late 1700s from a number of French scientists. Chemist Antoine-Laurent Lavoisier, like Boyle, was a meticulous experimenter. When investigating chemical reactions, he carefully weighed the reactants (the starting substances) and the products (the substances present after the reaction took place). Lavoisier discovered that when he burned small amounts of phosphorus and sulfur, the products of these reactions weighed more than the reactants!

Through careful measurements, Lavoisier showed that the weight gained by the products in these reactions was lost from the air. Based on this and other experiments, Lavoisier established the law of conservation of mass. It states that mass is neither lost nor gained during a chemical reaction. In other words, although matter may undergo changes during a chemical reaction, the quantity of matter (the total mass) is the same at the end of the reaction as it was at the beginning.

In 1794 French chemist Joseph-Louis Proust began experimenting with a compound called copper carbonate. Proust was able to chemically break down copper carbonate into its three components: the elements copper, carbon, and oxygen. During the next nine years, Proust analyzed hundreds of samples of copper carbonate. He discovered that these three elements were *always* found in the same ratio, by weight, in the compound. There were five parts copper to four parts oxygen to one part carbon—a constant ratio of 5:4:1.

Proust analyzed other compounds. He found that they, too, contained various elements that were always present in definite, or fixed, proportions by weight. These results led Proust to propose the law of definite proportions. It states that the ratio of elements in a given chemical compound is constant.

Here was compelling evidence for atomism. Assuming that elements were composed of atoms, it made perfect sense to Proust to suppose that atoms of one element would combine in a specific, fixed way with atoms of other elements to form compounds.

Proust also discovered that the proportions of elements in compounds were always whole numbers, such as 1:2 or 2:3. This finding seemed to indicate that atoms were discrete units that couldn't be divided—just as Democritus had suggested thousands of years earlier.

DALTON'S DISCOVERIES

British chemist John Dalton wondered how given quantities of different elements combined to form compounds. He calculated the relative weights of many elements—which led him to create the first "table of the elements"—and proposed the first atomic formulas for several compounds.

In 1803 Dalton published the first true atomic theory. It included what he believed to be fundamental truths about atoms.

- All matter is composed of extremely small particles called atoms that are indivisible and indestructible.
- All atoms of a given element are identical. They are the same size, have the same atomic weight, and possess the same chemical properties.
- All atoms of one element are different from those of any other element.
- Atoms cannot be created, divided into smaller particles, or destroyed.
- Different atoms combine in simple, whole-number ratios to form compounds.
- In a chemical reaction, atoms are never destroyed. They are simply separated, combined, or rearranged to form new molecules.

Dalton published a polished version of his theory in 1808, in the book *New System of Chemical Philosophy*. Dalton's atomic theory, coupled with a growing understanding of the nature of chemical substances and chemical reactions, provided the foundation for further investigations into the properties of elements and, ultimately, into the atom itself.

CHAPTER 2

ELECTRICITY AND THE DISCOVERY OF THE ELECTRON

In the early 1800s, scientists discovered a link between electricity and chemical substances. The source of electricity they used to make this discovery was a new device. It was first called a voltaic pile and, later, a battery. A voltaic pile consisted of a stack of metal disks sandwiched between pieces of wet cardboard. Two wires sprouted from the top of the device. When scientists set the wires into water in which salts or certain other substances were dissolved, the solution conducted electricity. That is, an electric current flowed from one wire through the solution to the other wire, forming a complete circuit.

In London, England, chemist William Nicholson and surgeon Anthony Carlisle inserted the wires from a voltaic pile into a container filled with water. To their surprise, they noticed bubbles of gas forming around the ends of the wires as the electricity flowed from the end of one wire, through the liquid, to the other wire.

ELECTRICITY As far back as the ancient Greeks, people had noticed that when substances such as amber (fossilized tree sap) were rubbed with a cloth, they would attract lightweight objects, such as hair or dust. By the middle of the 1600s, people had come to call substances such as amber "electrics." Small objects that were attracted to electrics were said to have become electrified. Some people believed that two kinds of electricity existed—one that caused objects to be attracted to one another and another that caused objects to repel one another.

In the 1740s, American statesman and inventor Benjamin Franklin showed experimentally that electricity was one thing that could exist in two different states. Things that became electrified could carry either a positive or a negative charge. The discovery that objects could acquire such charges was to have great implications for understanding atomic structure.

Nicholson and Carlisle collected the bubbles and analyzed them. Those forming on one wire were hydrogen. Those on the other were oxygen. Hydrogen and oxygen are the two elements that make up molecules of water. Nicholson and Carlisle had just discovered electrolysis. It separates out the elements in a chemical compound by passing electricity through a solution of that compound.

FARADAY'S FINDINGS

British chemist Humphry Davy lost no time in exploring electrolysis. Within months of its discovery, he was using the technique to separate out the elements from all sorts of chemical compounds. Davy discovered several new elements, including potassium and sodium. He realized that electricity, molecules, and atoms were somehow related. Davy hypothesized that molecules were made up of atoms and were held together by electrical forces.

British chemist and physicist Michael Faraday began his scientific career as Davy's assistant. In 1820 Faraday learned about several discoveries. A Danish chemist found that the needle of a compass (which is affected by magnetism) will turn toward a wire that has an electric current running through it. That same year, two French physicists showed that wires carrying an electric current could attract iron filings, just like a magnet did. Furthermore, depending on the direction of the current, the wires would either attract or repel each other, just as the two poles of a magnet did.

Faraday conducted a series of experiments to explore the link between electricity and magnetism. In 1831 he showed that an electric current could be generated by moving a magnet back and forth through a coil of wire. This experiment led to the development of the electric generator and, later, the electric motor.

Faraday proposed a radical hypothesis. He suggested that "lines of force" extend out from magnets and electrically charged objects. You can see magnetic lines of force for yourself when you set a sheet of paper over a magnet and scatter iron filings on top of the paper. Faraday theorized that lines of force formed a "field" around magnets and electrically charged objects. Those lines of force, he believed, were involved in attraction and repulsion. This proposal gave rise to the idea of the electromagnetic field, which is a combination of both electric and magnetic fields. Faraday's ideas on electromagnetic fields and forces lay important groundwork as scientists struggled to figure out what holds atoms together and how they interact.

In 1832 Faraday turned his attention to electrolysis. He soon found electromagnetic forces at work. In a solution of dissolved compound, there are charged particles (known as ions). Faraday proposed that during electrolysis, positively charged particles, or cations, move to the negatively charged cathode. Negatively charged particles, or anions, move to the positively charged anode. By deducing what was happening during electrolysis, Faraday provided another critical link between electricity, electromagnetic forces, and atoms.

ELECTROLYSIS TERMS Faraday coined a number of new terms to describe electrolysis. The solution or molten material through which the electric current passed (electrolysis doesn't work with a solid) was called the electrolyte. The components that actually make contact with the electrolyte (such as the tips of the wires) are called electrodes. The electrode that is attached to the negative pole of the power source (in Faraday's case, a battery) is called the cathode. The electrode attached to the positive pole of the power source is called the anode.

Although he published many papers about his discoveries, Faraday was not trained in mathematics. He was unable to support his conclusions with mathematical equations. As a result, other scientists did not immediately recognize the importance of Faraday's ideas.

In the 1860s, Scottish physicist James Clerk Maxwell studied Faraday's work. Maxwell developed four equations that expressed Faraday's ideas about electric and magnetic fields. This work led Maxwell to conclude that by themselves, electric and magnetic fields are static (unmoving). But when they interact, they form electromagnetic waves that travel through space. These energy waves (also called electromagnetic radiation) can be described by their wavelengths (the distance from one wave crest to the next), how much energy they contain, and their frequency (the number of wavelengths that pass a point in a given amount of time).

Maxwell used his equations to show that visible light is a form of electromagnetic radiation. In addition, he deduced that other forms of electromagnetic radiation, which are not visible, must also exist. In 1888 German physicist Heinrich Hertz discovered radio waves—a form of electromagnetic radiation with wavelengths too long for human eyes to detect. This finding confirmed Maxwell's hypothesis. In the closing years of the nineteenth century, scientists discovered several other kinds of "invisible" electromagnetic radiation. Although scientists didn't immediately realize it, these unseen waves were important clues to atomic structure.

THE ELECTROMAGNETIC SPECTRUM

Scientists have arranged all the known types of electromagnetic radiation into a sort of scale called a spectrum. At one end of the electromagnetic spectrum are short wavelength, high-frequency, high-energy electromagnetic waves. At the other end are long wavelength, low-frequency, low-energy waves. Visible light forms only a fraction of the total spectrum, sandwiched roughly in the middle. In order of increasing frequency (and energy), the types of electromagnetic waves making up the electromagnetic spectrum are radio waves, microwaves, infrared radiation, visible light, ultraviolet radiation, X-rays, and gamma rays.

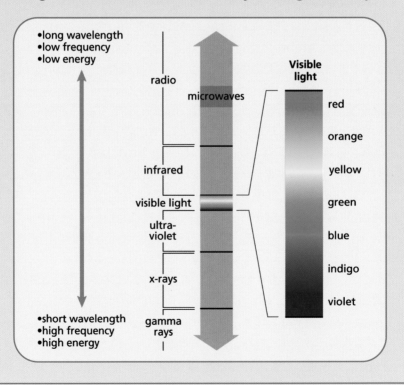

Cathode Ray Tubes

At roughly the same time that Maxwell was publishing papers about electromagnetic fields and waves, other researchers were studying electricity directly. For this research, they needed to see electricity flowing through empty space—a vacuum. In a vacuum, no atoms or molecules would get in the way of the electricity or interfere with their observations.

Vacuum pumps that could remove most of the air from a container had been around since the 1700s. In 1855 German glass blower Johann Heinrich Wilhelm Geissler invented an air pump that removed 99.9 percent of the air from glass tubes that he made himself. These Geissler tubes also had pieces of metal sealed into opposite ends that could serve as electrodes.

When scientists connected the positive anode and negative cathode of a Geissler tube to wires leading to a battery, a greenish glow appeared near the tube's cathode end. Was this glow electricity? Scientists weren't sure. If a magnet was moved back and forth beside the tube, the glow changed position. If a small piece of metal was sealed inside the tube directly in front of the cathode, it cast a shadow on the glow. This seemed to suggest that the glow was made up of rays, moving in straight lines, that were streaming out of the cathode. The rays came to be called cathode rays. Geissler tubes (along with later variations) were called cathode ray tubes.

Until the late 1890s, scientists were not sure what cathode rays actually were. But in 1897, British physicist Joseph John (J. J.) Thomson suggested that the rays were

Cathode ray tube

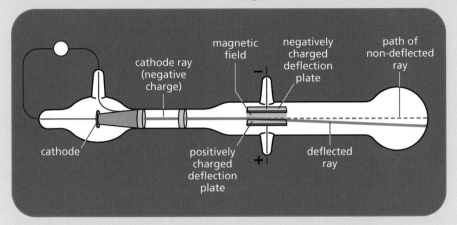

J. J. Thomson added deflection plates to a cathode ray tube to prove that cathode rays were charged particles. Since like charges repel, the negatively charged plate pushed away the negatively charged cathode rays. These negatively charged particles soon became known as electrons.

streams of particles that were much smaller than atoms. In fact, he believed they were *tiny pieces* of atoms. And he set out to prove it.

Thomson allowed cathode rays to speed through a cathode ray tube surrounded by electrically charged plates. The magnetic field created by the plates caused the cathode rays to curve away from the negatively charged plate and toward the positively charged one. Because charged particles are attracted to objects that carry the opposite charge, Thomson concluded that cathode rays were made up of negatively charged particles. Additional experiments allowed Thomson to calculate the mass of a single cathode ray particle. (He determined other variables, such as the strength of the magnetic field and the velocity of the particles, that he used in a formula to figure out the mass.) The particle was far smaller than a hydrogen atom—just 1/1837th of hydrogen's mass. The first subatomic particle had been found.

MENDELEYEV'S MASTERPIECE

The periodic table of the elements is a common sight in chemistry classrooms. This chart shows all the known elements organized in rows and columns. The table's organization changed through the centuries as scientists learned more about atomic structure.

In the mid-1800s, electrolysis continued to reveal the existence of new elements. Each element was different. Each had a unique atomic weight and distinct physical and chemical properties. Some, such as hydrogen, were gases. (Hydrogen was determined to be the lightest element with a mass of 1.00794 atomic mass units.) Other elements were metals or minerals. Some combined easily with other atoms to form molecules. Others tended to combine with nothing else.

In 1870 Russian chemist Dmitry Ivanovich Mendeleyev proposed a way to bring order to the disordered collection of elements. At that time, there were sixty-three known elements. Mendeleyev proposed a periodic table of the elements, in which all the known elements were organized by increasing atomic weight (in rows). The elements were also grouped together by characteristic chemical properties (in columns). He found that in setting up his rows and columns, he had to leave three gaps in the table in order for the known elements to fall into their proper places. Mendeleyev predicted that these gaps represented elements yet to be discovered. He was correct. The missing elements were all discovered within the next century.

For centuries scientists thought atoms were the smallest fundamental units of matter. Thomson's discovery shattered that notion. Thomson called the new particles corpuscles. This name was soon changed to electrons.

Thomson proposed a model of the atom that took the electron into account. Scientists generally accepted that normal atoms were electrically neutral. That meant that if one component of an atom—the electron—was negatively charged, then some sort of positively charged component should also be present to balance out the charges. Thomson suggested that an atom was a spherical cloud of positive charge in which tiny, negatively charged electrons were embedded, a bit like seeds in a watermelon. However, this model of the atom—dubbed the plum pudding model in Britain—enjoyed only a brief moment in the scientific limelight. Thomson would go down in history as the discoverer of the electron. But his atomic model had problems from the start.

REPORTING THE UNBELIEVABLE When J. J. Thomson reported that he had found the first subatomic particle, other scientists were skeptical. "At first there were very few who believed in the existence of these bodies smaller than atoms," Thomson recalled. "I was even told long afterwards by a distinguished physicist who had been present at my lecture at the Royal Institution that he thought I had been 'pulling their legs.'"

CHAPTER 3

Rays, Radioactivity, and Quantum Theory

German physicist Philipp Lenard questioned Thomson's plum pudding model of the atom. In the early 1900s, Lenard devised a cathode ray tube that allowed him to aim electrons at other objects. When directed at thin sheets of metal foil, the electrons passed straight through. If atoms (such as the atoms making up the sheets of foil) were as solid as Thomson's model suggested, streams of electrons shouldn't have been able to penetrate the foil.

In another experiment, Lenard shined ultraviolet (UV) light onto sheets of metal. Doing this caused electrons to be emitted from the metal. If electrons were embedded in atoms like raisins in a cake, they shouldn't have been so easy to dislodge.

In response to critics of his atomic model, Thomson suggested that perhaps electrons were only loosely embedded in the rest of the atom. This would explain Lenard's observations about electrons emitted from metals exposed to UV light. It would also explain the formation

Thomson's atomic model

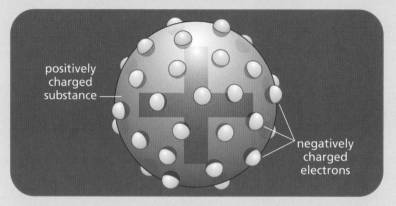

positively
charged
substance

negatively
charged
electrons

**After discovering the electron, an extremely small
negatively charged particle, J. J. Thomson proposed
that atoms consisted of electrons embedded in something that held a positive charge. This model is known
as the plum pudding model of the atom.**

of ions. Ions are atoms or molecules with negative or positive charges. Certain substances form ions easily. Take
table salt, NaCl, as an example. It is a compound with
molecules made up of one sodium (Na) atom and one
chlorine (Cl) atom. Dissolve salt in water, however, and
the salt molecules "come apart" (scientists say "dissoci-
ate") to form ions: positively charged Na^+ atoms and negatively charged Cl^- atoms. Thomson suggested that
electrons were the key to the formation of ions. If an atom
(or molecule) lost an electron, he reasoned, it would become a positively charged ion. If it picked up an extra
electron, it would become a negatively charged ion.

The fact that streams of electrons could pass through
metal foil—without bouncing back or off to the side—was
more difficult to explain. If atoms were essentially solid
spheres, why didn't they stop speeding electrons?

Lenard suggested that atoms were not solid spheres but collections of electrons, each of which was paired with a positive particle of some sort. But if that were the case, other scientists countered, why weren't there streams of these positive particles in a cathode ray tube? Why weren't anode rays coming from the tube's positive end?

The clues that would eventually provide an answer to these questions—and solve the puzzle of the atom's positive charge—had already been found. They had emerged from experiments conducted at roughly the same time as Thomson's discovery of the electron.

ROENTGEN'S NEW RAYS

In 1895 German physicist Wilhelm Conrad Roentgen had been experimenting with cathode rays. Roentgen wanted to learn how the rays affected various chemicals. Specifically, he was interested in whether cathode rays could cause chemicals to glow, or fluoresce. Fluorescence occurs when chemical substances gain energy—for example, after being exposed to sunlight—and then release the energy they've gained by glowing with a faint but distinctive light.

In his experiments, Roentgen had discovered that when cathode rays struck certain chemicals, especially a particular compound containing the element barium, they fluoresced. To study this effect in more detail, Roentgen coated sheets of paper with the barium compound and exposed them to cathode rays. Because the fluorescent glow emitted by the paper was faint, he covered the windows in his laboratory and enclosed the

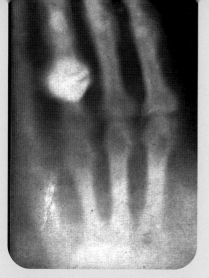

Wilhelm Roentgen took this X-ray photograph of his wife's hand in 1895.

cathode ray tube with black cardboard. In this dim environment, he hoped to see the fluorescence more clearly.

One day when Roentgen turned on the current to the cathode ray tube, he caught a faint flash of light out of the corner of his eye. The source of the light was one of his sheets of barium-coated paper. But it was not in the path of the cathode rays. It was clear across the room—and it was glowing.

Roentgen flicked off the current to the cathode ray tube. The paper across the room darkened. He flipped the current on again. The paper glowed. Intrigued, Roentgen passed his hand in front of the end of the cathode ray tube. He was amazed to see an image that showed the bones of his fingers projected on the paper.

Roentgen carried the paper into another room. When the cathode ray tube was on, the paper still glowed, even though a wall separated the paper and tube. What could explain this? Roentgen knew cathode rays weren't producing the glow. They did not extend beyond the glass of the cathode ray tube. The glow had to be produced by some other type of invisible rays. Not knowing what these rays were, Roentgen called them X-rays (*X* for "unknown"). X-rays were a new type of electromagnetic radiation, with considerably more energy than visible light or UV rays.

French physicist Antoine-Henri Becquerel learned about X-rays and wondered if they were produced by things other than glowing cathode ray tubes. For instance, did fluorescent minerals (which Becquerel often worked with in his laboratory) emit X-rays along with visible light?

Becquerel wrapped a photographic plate in black paper. The plate was covered with photographic emulsion. This silver-containing coating turned dark wherever it was struck by light. (Photographic plates came before photographic film. Currently, film cameras are being replaced by digital cameras.) Becquerel placed crystals of potassium uranyl sulfate—a fluorescent compound that contains the element uranium—on top of the wrapped photographic plate. He planned to set the crystals and the plate in the sun to "activate" the crystals so they would fluoresce. Becquerel knew that sunlight couldn't penetrate the paper to darken the plate. Neither would any fluorescent light the crystals gave off. But if the crystals gave off X-rays, these should penetrate the black paper and darken the photographic plate.

On the next clear day, Becquerel placed his crystal-topped, tightly wrapped photographic plate on a sunny windowsill. After a few hours, he brought the plate to a darkroom, unwrapped the plate, and developed it. The plate had turned a foggy dark grey beneath the crystals. Some type of radiation, other than visible light, had darkened its photographic emulsion.

Becquerel prepared to repeat the experiment with another photographic plate. But the next several days were cloudy. Unable to continue, Becquerel put the paper-wrapped plate, with the uranium-containing crystals on top, in a drawer. He then grew impatient. He decided to develop the plate just to make sure nothing was getting through the black paper. To his amazement, the plate was strongly darkened. Becquerel concluded that the crystals of potassium uranyl sulfate sitting on top of the plate had been giving off some powerful X-ray-like radiation that did not depend on sunlight and did not involve fluorescence. He guessed that the source of this radiation was the uranium atoms in the crystals.

RESEARCH ON RADIOACTIVITY

By 1898 another French physicist, Marie Curie, had discovered a great deal about the high-energy radiation given off by uranium atoms. She gave this phenomenon the name radioactivity. Marie Curie and her husband, Pierre, also discovered three other radioactive elements: thorium, polonium, and radium. Like uranium, all three elements gave off invisible but penetrating high-energy rays.

What was the source of the powerful radiation emitted by radioactive elements? Did it come in one form or many? The second question was answered quickly. The first took much longer. Experiments showed that radiation given off by uranium and other radioactive elements was composed of three different kinds of rays. Scientists called them gamma rays, beta rays, and alpha rays.

Marie and her husband and scientific partner, Pierre Curie, had to obtain and prepare their own samples of radioactive elements for their experiments. In 1898 they discovered the new element radium in ore tailings left over from uranium mining. Extracting radium to use in experiments meant crushing ore samples by hand and chemically separating out their components. It took the Curies twelve years—and hundreds of pounds of ore—to isolate a sample of pure radium. The work was exhausting and, although the Curies did not know it, very dangerous. No one at the time understood the hazards of radiation exposure. One of the most serious is an increased risk of cancer. In 1934 Marie Curie died of a blood disease typically caused by prolonged exposure to high-energy radiation.

Gamma rays were similar to X-rays, although even more powerful. Beta rays were quickly recognized as fast-moving streams of electrons. (Beta rays and cathode rays were thus shown to be the same thing.) Electrons given off by radioactive substances became known as beta particles. In 1903 New Zealand physicist Ernest Rutherford discovered that alpha rays were streams of fast-moving *positively* charged particles. Why and how high-energy radiation and fast-moving particles were being released from radioactive elements was still unclear. But while experimental scientists such as Rutherford kept looking for answers in the laboratory, another group of physicists took a new approach.

THOUGHTFUL SOLUTIONS

At the end of the nineteenth century, a number of physicists began trying to answer questions about energy, radiation, and atomic structure without conducting experiments. Instead, they thought through problems, developed hypotheses about them, and then attempted to prove these hypotheses using math. These scientists were the first theoretical physicists.

German theoretical physicist Max Planck wanted to know more about electromagnetic radiation. Nearly all the different kinds of radiation making up the electromagnetic spectrum were known by that time. Most scientists believed that all types of electromagnetic radiation flowed continuously as they moved through space. Planck proposed that electromagnetic radiation traveled in discrete units, in measurable "packets" of energy. He called these packets of energy quanta (just one is a quantum). Planck showed mathematically that the amount of energy in a given quantum of electromagnetic radiation is related to its wavelength. The shorter the wavelength (and therefore higher the frequency) of any electromagnetic radiation, the higher the energy content of its quanta. Planck introduced this quantum theory in 1900.

In 1905 German physicist Albert Einstein—one of the greatest scientists of the twentieth century—pondered Planck's idea about energy quanta. Theoretically, he went beyond Planck's work and proposed a new way of thinking about visible light. (Remember that visible light is electromagnetic radiation with a certain range of wavelengths.) Until this time, scientists believed that light travels only in waves. Einstein argued that light can also act as though it were made up of particles of energy. He called the particles light quanta. (The name was later changed to photons.) This proposal was revolutionary. Many science historians mark the introduction of quantum theory and Einstein's ideas about the particle nature of light as the beginning of modern physics.

Einstein's theory of light quanta explained an observation that had stumped physicists of his day: the photoelectric effect. In this phenomenon (which Lenard is credited with discovering), light shining on metal will cause electrons to be ejected from the metal's surface—but *only* when the frequency of light exceeds a certain value. In other words, only light of certain wavelengths—and so with a certain amount of energy—can cause the photoelectric effect.

In explaining the photoelectric effect, Einstein assumed that light was not only *radiated* in quanta but also *absorbed* in quanta. When light struck a surface, he reasoned, it was absorbed by electrons bound to the atoms making up that surface, one quantum at a time. If the energy content of light quanta striking the surface was below a certain amount, electrons wouldn't be affected. But if that quantum energy was sufficient to overcome the forces holding electrons in place in the atom, the electrons would be set free.

Einstein's ideas about photons and quanta so neatly explained the photoelectric effect that many physicists who had been reluctant to accept quantum theory began to embrace it. From 1905 onward, the study of atomic structure proceeded with the new understanding of light, energy and, ultimately, matter that quantum theory provided.

CHAPTER 4

NUCLEAR ATOMS AND THE BOHR MODEL

In 1906 Rutherford was still experimenting with alpha rays. He determined that alpha rays were streams of fast-moving, positively charged particles. Alpha particles had roughly eight thousand times the mass of an electron. Rutherford wondered what might happen if he fired alpha rays at atoms. Perhaps the way in which alpha particles behaved might reveal clues about atomic structure.

Rutherford designed an experiment in which streams of alpha particles coming from samples of the radioactive element radium bombarded thin sheets of gold foil. Behind the foil was a sort of photographic device called a scintillation screen. As alpha particles passed through the foil, they struck the screen. How they struck the screen revealed how they traveled—in straight lines, unaffected by anything in their path, or in some other fashion.

Rutherford asked two of his graduate students, Johannes Geiger and Ernest Marsden, to carry out this experiment. By 1909 Geiger and Marsden had run the test

Gold Foil Experiment

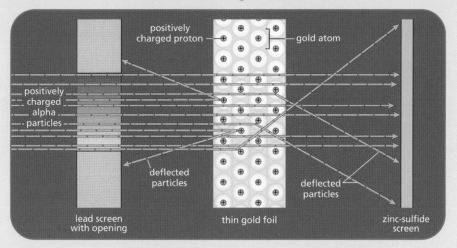

positively
charged proton — gold atom

positively
charged
alpha
particles

deflected
particles

deflected
particles

lead screen
with opening

thin gold foil

zinc-sulfide
screen

Scientists discovered protons with this experiment. They shot a stream of positively charged alpha particles at a sheet of thin gold foil. Most alpha particles went straight through the foil, but a small number were deflected. Positively charged protons within the gold atoms repelled the alpha particles, causing the deflections.

many times. When they fired alpha particles at the gold foil, nearly all of them passed through it as if it wasn't there. The particles struck the scintillation screen behind the foil, leaving a fogged area that showed these particles had never altered from their original path.

But approximately one in eight thousand alpha particles had not gone straight through the foil. For some reason, they had changed course and headed off at an angle. Some appeared to have even bounced backward!

Rutherford was delighted. The results were a bit like firing bullets at a sheet of tissue paper and having some of them bounce back. It seemed impossible. But Rutherford knew it could mean only one thing: the alpha particles with the altered paths had encountered something as they traveled through the foil. Whatever they encountered had to be at least as heavy as they were.

Closer analysis of the patterns on the scintillation screens revealed that the alpha particles with the altered paths hadn't actually bounced off anything. They had *turned away* from something before they struck it. Since alpha particles carry a positive charge, Rutherford deduced that the particles had been deflected by something inside the atom that also carried a positive charge.

In 1911 Rutherford suggested a new model of the atom. He proposed that an atom's positive charges—and nearly all its mass—were concentrated in a tiny atomic nucleus in its center. Spinning around the nucleus were an atom's electrons, somewhat like planets circling the Sun. The attraction between the negative electrons and the positive nucleus held the atom together.

In discovering the nucleus, Rutherford realized that the simplest nucleus belonged to the smallest and lightest atom: hydrogen. He called the hydrogen nucleus the proton. Rutherford speculated that larger atomic nuclei might contain two or more protons.

The nuclear atom explained some observations about atoms. For instance, with electrons circling the atomic nucleus, it seemed reasonable to assume they could be easily lost or gained to form ions. But the nuclear atom also raised new questions. What kept negative electrons in place around a positive nucleus? Since opposite charges attract, why didn't the electrons get pulled into the nucleus? Furthermore, it was an established fact that when objects move in a circular motion, they give off electromagnetic radiation. But electrons didn't. Why not?

BOHR'S MODEL

Niels Bohr, a Danish theoretical physicist working in Rutherford's lab, tackled these questions of electrons and atomic structure. To begin with, he knew that when an element is heated, it releases energy. And when each element releases energy, it gives off only certain wavelengths of light. These wavelengths can be viewed as spectral lines—a pattern of colored lines separated by dark spaces. Each element's pattern of spectral lines is unique, like a fingerprint.

With spectral lines in mind, Bohr began to think about electrons, orbits, and wavelengths of energy in the context of quantum theory. He focused on the simplest atom, hydrogen. Hydrogen was believed to have one proton and one electron. Its pattern of spectral lines was well known. The wavelengths had been precisely measured.

Bohr hypothesized that the electron in the hydrogen atom could take on a number of different orbits, at different distances from the nucleus. As long at it remained in a particular orbit, the electron wouldn't gain or lose energy. But if it changed orbits, a change in energy should take place. Bohr suggested that if an electron gained energy, it would move into a higher energy orbit, farther from the nucleus. If it lost energy, it would slip back into a lower energy orbit, closer to the nucleus.

Bohr recognized that the idea of electrons jumping between orbits fit well with the idea of light quanta (photons) in quantum theory. Perhaps an atom could absorb (and emit) only photons of certain sizes. When an electron jumped from one orbit to another, the amount of energy gained or released would be equal to a single photon of a characteristic wavelength. This would explain the distinctive spectral lines of each element.

Bohr mathematically defined a set of "permitted orbits" for the hydrogen electron. The orbits allowed for the gain or release of quanta of certain energies (and so certain wavelengths of light). As it turned out, those wavelengths matched hydrogen's spectral lines.

In coming up with the mathematical formula that described electron orbits, Bohr had to use whole numbers for one of the terms. This was because quanta were involved, and it is not possible to have a fraction of a quantum. The number Bohr used, represented by the letter n, came to be called the principal quantum number. It represents the size of orbits in an atom.

Bohr incorporated his ideas and calculations into what became known as the Bohr model of the hydrogen atom. The Bohr model had one electron that orbited a nucleus with a positive charge. The electron had a series of possible orbits it could follow around the nucleus. The model proposed that when a hydrogen atom was in its normal state, the electron would be in the orbit closest to the nucleus. If the electron absorbed a photon of sufficient energy, however, it would jump up to the next higher orbit. When it jumped down, the atom would emit a photon. In 1913 Bohr published his ideas and introduced his model to the scientific world.

The Bohr model received mixed reviews. One problem was that it modeled only the simplest of atoms—hydrogen. Yet Bohr seemed to be on the right track.

Atomic Numbers and the Periodic Table

Bohr's model drew attention to the charge on an atom's nucleus. His calculations as to which orbits were allowable in an atom took into account the positive charge on the nucleus.

Henry Moseley was a student of Rutherford's doing research at the Cambridge University's Cavendish Laboratory in Cambridge, England. Moseley was intrigued by the link between the amount of energy that could be absorbed or released by atoms and the charge on their nuclei. In 1913 Moseley measured the wavelength of X-rays emitted by different elements, using a technique called X-ray diffraction. Moseley realized that if Bohr's model of the atom was correct, the energy of X-rays emitted by different elements should give a direct measurement of the amount of positive charge in their atomic nuclei.

Moseley's hypothesis was correct. He tested all the known elements, determining the size of their nuclear charge, which he called the atomic number. He assigned hydrogen an atomic number of 1, helium 2, and so on up to uranium with an atomic number of 92.

Finding the atomic numbers of the elements led to a slight reorganization of the periodic table. Determining the known elements' atomic numbers also revealed seven gaps in the table, representing elements yet to be discovered between hydrogen and uranium.

AIMING AT THE NUCLEUS

In August 1914, world events interrupted scientific research. World War I (1914–1918) broke out in Europe. Many young scientists went to join in the fighting.

By 1917 Ernest Rutherford was one of the few scientists left in his laboratory. He began a new set of experiments that involved bombarding air (which is largely oxygen and nitrogen) with alpha particles. During these experiments, Rutherford accomplished the first nuclear disintegration, or breaking apart of an atomic nucleus. With alpha particles, he managed to knock a proton out of the nucleus of a nitrogen atom. This seemed to confirm that the nuclei of all atoms, not just hydrogen, were made up of protons.

After the war, scientists refocused their attention on atomic structure. Bohr's model worked well for hydrogen but not for atoms with more than one electron. The spectral lines of hydrogen had also been studied in more detail. It was found that each line was actually made up of finer lines. Bohr's model, it seemed, was too simple.

WAVE-PARTICLE DUALITY

In 1923 French physicist Louis de Broglie was inspired by Einstein's idea that light can behave as both a wave and a particle. He wondered why light should be unique. Perhaps *all* forms of matter exhibit both wave and particle characteristics. If this was true, electrons should behave like waves as well as particles.

De Broglie's radical idea about the wave-particle duality of matter was confirmed experimentally. All types of matter do exhibit wave characteristics, although the

SPIN In 1925 Dutch physicists George Eugene Uhlenbeck and Samuel Abraham Goudsmit discovered that electrons spin as they travel around the nucleus. The two researchers determined that electrons could spin in only two ways, "up" or "down." (These directions are arbitrary and have nothing to do with gravity.) In discovering this property of electron spin, it became obvious that electrons can behave like minute magnets. (Remember that any moving electric charge will produce a magnetic field.) Furthermore, it meant that electrons possess a characteristic that physicists call angular momentum. Angular momentum also applies to objects such as bicycle wheels. It basically measures how difficult it is to stop the object from spinning. Electrons were the first—but not the only—subatomic particles found to have spin.

waves are nearly impossible to detect in large objects. But in something as minute as an electron, the wave characteristics are obvious.

Furthermore, if an electron behaves like a wave strung out around the nucleus rather than like a particle spinning around it, then that electron would be restricted to orbits that are exactly the right size to contain a whole number of waves (the crests and the troughs). Instead of electrons "jumping" from one orbit to the next, de Broglie maintained, it's the frequency of an electron's wave that can change, gaining or releasing energy in the process.

SCHRODINGER'S PROBABILITIES

In 1926 Austrian physicist Erwin Schrodinger expanded

on de Broglie's wave ideas. He transformed them into a mathematical equation that describes the behavior of electrons and other tiny particles. Schrodinger's equation, as it is called, didn't just describe hydrogen atoms or a handful of atoms but *all* atoms.

Unlike the Bohr model, in which electrons follow strictly defined orbits, Schrodinger's model describes the *probability* of finding an electron at any given moment. It is most likely to be in an orbit predicted by the Bohr model. But it can be found anywhere within a region of space inside the atom known as an orbital. Orbitals look a bit like fuzzy clouds. They're zones in which a given electron might be found. They come in different shapes and sizes, depending on the atom and the number of electrons it contains.

Austrian-born U.S. physicist Wolfgang Pauli defined an electron's behavior within the atom even further. He used quantum theory to develop the Pauli exclusion principle. It states that no two electrons in the same atom can be in exactly the same place at the same time with the same spin. To develop this idea mathematically, Pauli incorporated three more quantum numbers to describe the wave pattern of an electron around the nucleus and thus the size and shape of its orbital. In addition to n, the other quantum numbers are s for electron spin, l for the electron's angular momentum as it revolves around the nucleus, and m for the orientation, or tilt of the electron waves with respect to one another. No two electrons within the same atom can have exactly the same four quantum numbers.

Rutherford's, Bohr's, and Schrodinger's Models

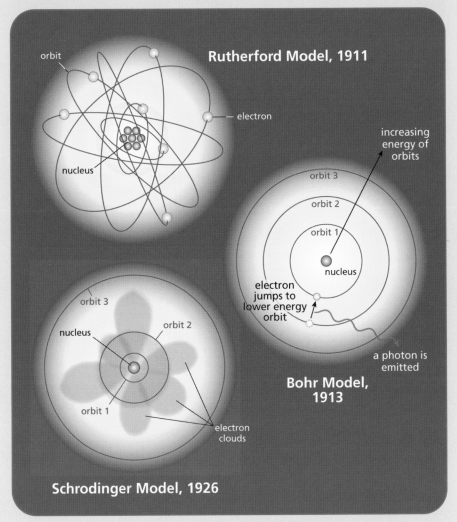

Rutherford Model, 1911

orbit

electron

nucleus

increasing energy of orbits

orbit 3

orbit 2

orbit 1

nucleus

electron jumps to lower energy orbit

a photon is emitted

Bohr Model, 1913

orbit 3

nucleus

orbit 2

orbit 1

electron clouds

Schrodinger Model, 1926

In 1911 Ernest Rutherford suggested that most of an atom's mass is in the positively charged nucleus and that electrons orbit the nucleus much like planets orbit the Sun. Two years later, Niels Bohr proposed a model of a hydrogen atom that allowed specific orbits for the electron, depending on how much energy it had. When the electron lost energy, which caused it to drop into an orbital closer to the nucleus, it emitted a photon. In 1926 Erwin Schrodinger developed an equation for all elements that described the mathematical probability of finding an electron in a given location. Atomic models based on his equation show "probability clouds" in which electrons are likely to be found.

UNDERSTANDING CHEMICAL REACTIONS

By the early 1930s, scientists' understanding of the atom had advanced a great deal. Using mathematics, quantum theory, and the exclusion principle, researchers could describe how electrons were distributed inside atoms—even inside atoms that have many electrons. Knowing this electron configuration helped answer many questions about why atoms behave as they do in chemical reactions.

Orbitals are grouped together in series called shells. Shells are numbered outward from the nucleus. Electrons fill shells that are closest to the nucleus first. (These shells have the lowest energy.) The first shell can hold two electrons. The second can hold eight. Shells three, four, and five can hold eighteen, thirty-two, and fifty electrons, respectively. Atoms with shells that are completely full tend to be nonreactive. That is, they do not combine readily with other atoms. Elements such as helium (with two electrons; first shell full) and neon (with ten electrons; shells one and two full) have this kind of electron configuration.

Atoms with outermost shells that are only partially full are much more reactive. Sodium has just one electron in its outermost shell. That lone electron can easily be pulled off by other atoms in a chemical reaction. An atom of fluorine, with seven electrons in its second shell, is also very reactive—it can easily gain an electron to complete its second shell.

In the first three decades of the twentieth century, quantum theory helped unlock many secrets about atomic structure. But scientists still had much more to learn.

CHAPTER 5

NEUTRONS, CLOUD CHAMBERS, AND COSMIC RAYS

In 1930 physicists had a new mystery to tackle. Irène Joliot-Curie (the daughter of Marie and Pierre Curie) and her husband, Frédéric Joliot-Curie, reported that when they bombarded the element beryllium with alpha particles, some kind of neutral radiation had been produced. This radiation was powerful. It knocked protons out of paraffin wax, which is rich in hydrogen. Curie and Joliot published a paper stating that they believed their newly discovered radiation was some type of gamma ray.

Nearly fifteen years earlier, Ernest Rutherford had speculated that the proton might not be the only nuclear particle. But he had been unable to find such a particle. James Chadwick, who worked with Rutherford, also believed another nuclear particle existed. When he read the Joliot-Curie paper, he realized the French scientists had stumbled upon a neutral component of the atomic nucleus. If he could find the same radiation and prove his suspicions, he'd crack one of the great unsolved mysteries about atoms.

Since the discovery of the atomic nucleus, physicists had thought that the nuclei of all atoms consisted of protons. But there was a problem with the atomic weights of the elements. Take helium, for example. With an atomic number of 2, it has two electrons and therefore should have two protons (two positive charges to balance the two negative charges of the electrons). But strangely, helium has a mass (atomic weight) of 4, not 2. Lithium, the next element in the table, has 3 electrons. It has a mass of nearly 7. The same sort of problem existed with all the other elements except hydrogen. Chadwick concluded that nuclei must contain another particle along with protons.

Chadwick's experiments confirmed his hunch. The neutral radiation coming off beryllium was made up of particles. Those particles had about the same mass as a proton but a neutral charge. They also had enough power to knock protons out of atomic nuclei. In 1932 Chadwick reported that he had discovered the neutron.

Once again, a new picture of the atom emerged. Atoms consisted of three particles: electrons, protons, and neutrons. With the exception of hydrogen, the atomic nuclei of every element contain neutrons in addition to protons.

Chadwick's discovery solved the atomic weight problem. Neutrons and protons together account for (almost) all of an atom's atomic weight. The discovery of the neutron also explained isotopes. Isotopes are different forms of an element with different numbers of neutrons. For example, most hydrogen atoms have no neutrons. But several isotopes of hydrogen do have them. A hydrogen atom with one neutron in its nucleus is called deuterium. Hydrogen atoms with two neutrons are the isotope known as tritium. Isotopes of the same element all have the same atomic number (equal to the number of protons) but different atomic weights (equal to the number of protons plus neutrons).

TRANSMUTATIONS AND ISOTOPES

Working with Ernest Rutherford, Frederick Soddy discovered that as radioactive elements break down, or decay, by emitting alpha particles and beta particles, they actually turn into other elements. The process by which one element changes into another is called transmutation. In a sense, it is nature's way of accomplishing what the ancient alchemists had tried to do.

Soddy later discovered that a radioactive element can come in different forms, each with a particular atomic weight. To describe these different forms of the same element, Soddy coined the term *isotope*, a word that means "at the same place." It refers to the fact that all isotopes of an element are located at the same place on the periodic table. Soddy later showed that nonradioactive elements also have isotopes.

ACCELERATING PROTONS

Also in 1932, two physicists working with Ernest Rutherford, John Douglas Cockcroft and Ernest T. S. Walton, designed and built the first proton accelerator. The device used a high-voltage electric charge to accelerate streams of protons to speeds high enough to smash apart atomic nuclei. In April 1932, Cockcroft and Walton used their accelerator to break apart the nucleus of a lithium

THE CLOUD CHAMBER

Have you ever seen vapor trails left behind in the sky by high-flying aircraft? The trails consist of water droplets that condense on particles of exhaust shot out of jet engines. The cloudlike trails provide a record of the plane's path through the air. The principle behind the cloud chamber was similar but on a much smaller scale. It was basically a large glass chamber containing chilled water vapor. When electrically charged particles sped through the chamber, they ionized air molecules. Vapor condensed on these ionized particles to create tiny clouds that formed along the tracks of the particles—like miniature vapor trails. When rigged up to a photographic device, the cloud chamber made it possible to record the tracks of speeding particles on film. An entire sequence of events could be recorded in the patterns of trails made by particles zooming through a cloud chamber, including the breakdown of one particle into another.

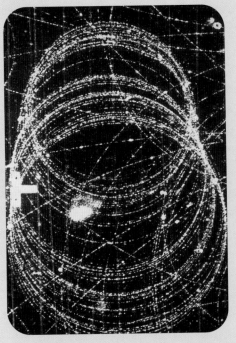

Electron and positron spiral tracks can be seen in this cloud chamber photograph.

atom (containing three protons and four neutrons). The disintegration was visible in a cloud chamber. Rutherford described it as one of the most beautiful things he'd ever seen.

Cockcroft and Walton's proton accelerator produced the first artificially accelerated particles that could disintegrate an atomic nucleus. But other scientists soon realized that they had a natural particle accelerator available in the form of cosmic rays.

Cosmic rays are high-energy particles from outer space that are constantly bombarding Earth's upper atmosphere. With the invention of cloud chambers, researchers could "catch" cosmic rays on film and record their tracks. Carl David Anderson, a physicist at the California Institute of Technology, began taking pictures of cosmic rays passing through a cloud chamber he had designed himself. In 1932 Anderson recorded the historic passage of a positively charged electron passing through the center of his cloud chamber. Other scientists were skeptical. But just a few months later, two other physicists confirmed Anderson's discovery using their own cloud chamber. This positive electron was given the name positron.

Where do positrons come from? Evidence was growing that they were produced by collisions of cosmic rays with atoms in the atmosphere. But there was more to the story. Pictures of particle tracks from cloud chambers showed other particles. Those particles were far more penetrating than electrons and positrons.

In 1937 Anderson and his graduate student Seth Neddermeyer found a particle that had a mass between that of a proton and an electron. They called it a

$E = mc^2$ **In 1905 Albert Einstein proposed more than just the idea of light quanta. He also introduced the world to his most famous equation, $E = mc^2$. The equation tells us that mass *(m)* is equivalent to energy *(E)*. It also tells us that the speed of light squared *(c^2)* relates the amount of energy in a given amount of mass and vice versa. What does this mean? At the scale of subatomic particles, mass can change into energy in the form of light, heat, or motion (kinetic energy). By the same token, energy can also turn into mass. Such transformations occur routinely in particle accelerators, where particles can be converted into energy and energy can become transformed into new particles.**

mesotron. (Meso is Greek for "middle.") The name mesotron was later changed to muon.

MORE—AND STRANGER— PARTICLES

World War II (1939–1945) brought a halt to much of the research on cosmic rays. Many atomic physicists took jobs with the military. Several dozen scientists in the United States helped develop the first atomic bombs. Atomic bombs release enormous amounts of energy through the splitting apart, or fission, of atomic nuclei. Scientists used a rare natural isotope of uranium (uranium-235) and an artificially created isotope of plutonium (plutonium-239) to create the two atomic bombs that the United States dropped on Japan in 1945.

Following the war, studies of cosmic rays and mysterious subatomic particles resumed. It soon became obvious that particles such as positrons and muons were secondary particles. They were created by powerful primary cosmic radiation striking the atmosphere at high altitudes. A number of cosmic rays scientists set up cloud chambers at mountain observatories.

Technological advances made during the war also enhanced cosmic ray research. New, very sensitive kinds of film had been invented. They were exposed directly to cosmic rays at mountaintop observatories or inside balloons or airplanes flown at high elevation. The new film yielded detailed photographs of particle trails. In 1947 the first example of a strange particle, the kaon, was discovered. Strange particles left odd V-shaped tracks on cloud chamber images. Also in 1947, a particle showed up that was slightly heavier than the muon. Japanese theoretical physicist Hideki Yukawa had predicted this particle's existence in 1935. It became known as the pion.

By the 1950s, researchers had a pretty good idea of what was happening in the upper atmosphere and why they were finding so many subatomic particles there. They proposed that atomic nuclei from outer space were colliding with atoms in the upper atmosphere. The fragments that resulted from these collisions were mostly protons, neutrons, and some small, lightweight nuclei. But there were also pions, which can carry positive or negative electrical charge or no charge at all. Uncharged pions apparently decayed into gamma rays. These, in turn, produced showers of electrons and positrons as they traveled farther down into Earth's atmosphere. Charged pions broke down to become muons. Some muons survived their journey through the atmosphere. It was suspected that they penetrated deep into Earth. Others decayed into electrons, accompanied by the release of an elusive particle called the neutrino (discovered in 1956), which could pass right through Earth.

Cosmic ray studies opened a door on a new world of weird and wonderful subatomic particles formed in the high-energy environment of the upper atmosphere. Was there more yet to come? Most particle physicists believed that they had just scratched the surface of what might exist in this subatomic world.

PAULI AND THE NEUTRINO

Neutrinos are small, neutral subatomic particles. Scientists believe they have almost no mass. The Sun and other stars produce enormous numbers of neutrinos. The decay of certain radioactive elements also produces neutrinos.

Wolfgang Pauli anticipated the existence of the neutrino in 1930. Pauli proposed the neutrino as a solution to a problem physicists encountered in trying to figure out beta decay. In this process, radioactive substances emit fast-moving beta particles (electrons). Scientists determined that a small amount of energy seemed to be lost in beta decay. (It is an established law of physics that energy cannot be created or destroyed.) Pauli proposed that the products of beta decay must include a lightweight, neutral particle that carried away this missing energy. He called this as yet undiscovered particle a neutrino.

CHAPTER 6

THE PARTICLE EXPLOSION

As physicists created more powerful particle accelerators, they continued searching the building blocks of atoms. The faster the particles travel in an accelerator (or anywhere), the more energy they gain. When the speeding particles in an accelerator have reached sufficient speed and energy, they are allowed to collide with other particles. In the collisions, all sorts of subatomic particles are released—and even created. Most break down into still other subatomic particles as quickly as they are formed. Special detectors record the outcome of these particle collisions and the faint traces that prove these bizarre collections of subatomic particles really do exist.

In 1955 scientists using a proton accelerator called the Bevatron at Lawrence Berkeley National Laboratory in Berkeley, California, discovered the antiproton. It was the second antiparticle. (The positive electron, or positron, was the first.) The antiproton is the twin of a proton, equal to it in mass but carrying a negative charge.

A trickle of new discoveries quickly turned into a flood. By the early 1960s, more than one hundred new kinds of subatomic particles had been discovered. Physicists started calling the collection the particle zoo. They had no real understanding of how all these particles were related or what forces controlled them.

ANTIMATTER For every known particle of matter, scientists have discovered a corresponding antiparticle, or, particle of antimatter. Antiparticles typically have a charge opposite to their matter partners. Otherwise, particle/antiparticle pairs share similar properties. If a particle meets its antiparticle, however, the two will annihilate each other, becoming instantly transformed into pure energy. (Collisions of matter and antimatter are rare, since antimatter is not found naturally on Earth, except briefly in tiny quantities.) Although it sounds too strange to be real, antimatter is as real as matter. In the mid-1990s, scientists successfully joined an antielectron, or positron, to an antiproton to form an antiatom of antihydrogen!

ENTER THE QUARK

The first suggestion of an underlying order came in 1962. Working independently, two theoretical physicists, American Murray Gell-Mann and Israeli Yuval Ne'eman, arranged groups of subatomic particles into symmetrical patterns, based on their electric charge and how much "strangeness" they had. One of Gell-Mann's patterns seemed to have a particle missing. Believing that it probably did exist and simply hadn't been discovered yet, Gell-Mann calculated the missing particle's properties, including its mass. In 1964 a subatomic particle named omega-minus was found. It had almost exactly the mass Gell-Mann had predicted.

STRANGENESS In 1953 Murray Gell-Mann solved a mystery concerning quantum theory and certain subatomic particles. Most particles found in particle accelerators broke down breathtakingly fast. But a few broke down much more slowly. What could be responsible for the difference? Gell-Mann proposed a new quantum property of particles to account for this behavior that he called strangeness.

In 1964, again working independently, Gell-Mann and George Zweig proposed another seemingly radical idea: all the newly found particles were not truly fundamental. Instead, they were built from even more basic building blocks. Gell-Mann called them quarks. Zweig called them aces. The name quark stuck.

Gell-Mann originally identified three types of quarks: up, down, and strange. (Some physicists started calling these quark characteristics flavors to distinguish them from other quark properties including spin and charge.) Each quark had an antiquark. Quarks also had charges. They weren't whole number charges such as +1 or −1. They were fractional charges of +1/3, −1/3, +2/3, or −2/3. According to Gell-Mann's quark model, a proton is made up of three quarks: two up and one down. Their fractional charges add

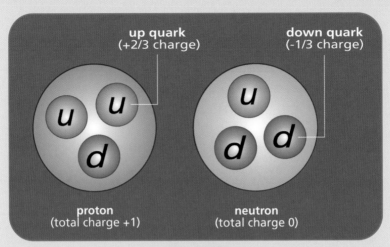

up quark
(+2/3 charge)

down quark
(-1/3 charge)

proton
(total charge +1)

neutron
(total charge 0)

Quarks

Scientists have determined that protons and neutrons are made of smaller particles known as quarks. Quarks have fractional charges, which when added together, give the correct charge for the particles they make up.

up to +1, which is known to be a proton's charge. A neutron is also made up of three quarks but, in this case, one up and two down. Their charges add up to 0, giving the neutron a neutral charge. Other, lighter-weight particles are made up of one quark and one antiquark.

At first, many physicists were skeptical. But as they started working with the model, they found that it could explain every subatomic particle that had been discovered up to that time. Still, no one had ever seen a quark. Did they really exist?

LUMPY EVIDENCE

Indirect evidence for the existence of quarks came in 1967. Scientists working at the Stanford Linear Accelerator fired electrons at protons. When they analyzed the scattering patterns of the electrons, it appeared that some of the electrons had bounced back from protons. And judging by the way they bounced, scientists concluded that there were three lumps inside each proton. Were those lumps the mysterious quarks? Experiments carried out at other accelerators gave similar results. It seemed the lumps might just be quarks after all.

Gell-Mann's model had originally proposed three quarks. But in 1974, scientists working at two different particle accelerators in the United States announced—on the same day—that they had found a new particle believed to contain a fourth flavor of quark. It was dubbed the charm quark. Later discoveries at other accelerators turned up other new particles shown to contain charm quarks too. At this point, the quark model became almost universally accepted in the world of particle physics.

Then in 1977, a team of scientists led by Leon Lederman at Chicago's Fermi National Accelerator Laboratory (Fermilab) discovered a fifth flavor of quarks. They called them bottom quarks. Chances were good, many scientists believed, that the bottom quark would also have a partner. Was a top quark still out there, waiting to be found? The search began almost immediately.

The top quark was finally discovered in 1995 at Fermilab, after an intensive effort involving more than 450 physicists. This discovery filled a hole in what has come to be the currently accepted model of atomic structure. It is a model that best represents our understanding of how atoms are built from subatomic particles and how those particles behave.

CHAPTER 7

THE STANDARD MODEL AND BEYOND

The Standard Model of Fundamental Particles and Interactions is also called the Standard Model Theory or simply the Standard Model. It represents the most current explanation of the nature of matter. It describes matter in terms of three basic kinds of elementary particles: quarks, leptons, and force carriers. Everything in the universe is believed to be made from these building blocks. The illustration on the next page shows these fundamental particles organized in a grid. Keep in mind, however, that for every particle that makes up matter, there is a corresponding antiparticle that makes up antimatter. To make the illustration easier to understand, these antiparticles are not included.

QUARKS, HADRONS, AND LEPTONS

Let's begin with quarks. Quarks come in six flavors. They are typically grouped into three sets: up/down, strange/charm, and top/bottom. (Top and bottom quarks

Elementary Particles

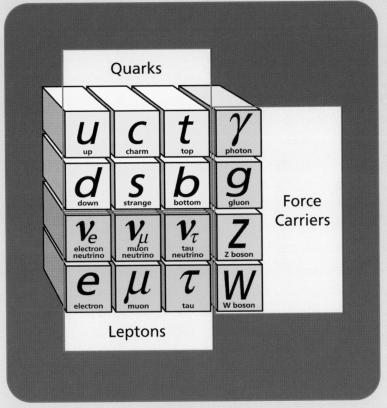

This chart sums up physicists' current understanding of the known fundamental particles in the universe.

are sometimes called truth and beauty.) Quarks carry fractional charges. Up, charm, and top quarks carry a +2/3 charge. Down, strange, and bottom quarks carry a −1/3 charge. All quarks have a −1/2 spin. Furthermore, they also have color (see page 61).

Quarks are never found alone. They exist only in groups that together form composite particles called hadrons. Two kinds of hadrons exist: baryons and mesons. Baryons each contain a combination of three quarks. Protons, for example, are baryons formed from two up quarks and one down quark. Neutrons, also baryons, each contain one up quark and two down quarks. Baryons always have whole integer charges, such as 0 for neutrons and +1 for protons, because they are formed from quarks whose fractional charges always add up to a whole number.

Mesons are the second type of hadrons. These particles are formed from one quark plus one antiquark. Examples of mesons include the pion, the kaon, and the psi meson. Because mesons are formed from quarks and antiquarks, they tend to be unstable and break down quickly.

This brings us to the leptons. Unlike quarks, which always form composite particles, leptons are solitary. The best-known lepton is the electron, which carries a negative charge. Two other charged leptons are the muon and the tau. They both have a greater mass than the electron. They decay quickly and are not found in ordinary matter. The remaining leptons are the lightweight neutrinos. There is a neutrino to correspond to each of the charged leptons. In other words, there is an electron neutrino, a muon neutrino, and a tau neutrino. All three kinds of neutrinos are elusive. They are so small and so nonreactive that they do not interact with other kinds of matter. In fact, neutrinos are passing right through your body right now.

> **QUARK COLOR** In the Standard Model, quarks have a property called color. It is necessary because of the Pauli exclusion principle, which states that no two identical particles can occupy exactly the same quantum state. Take a proton, for example. It is composed of two up quarks and one down quark. Since those two up quarks cannot be absolutely identical, the idea of color was introduced to give quarks a further distinguishing characteristic. Quarks are said to come in three colors: red, green, and blue. (The colors, of course, are not real but simply define three distinct quantum states.) So although a proton contains two up quarks, those two quarks are different colors and the exclusion principle is not violated.

FORCE CARRIERS

Leptons and quarks are thus the building blocks of all matter. But what holds them together in the various particles and atoms they form? The answer is four fundamental forces: electromagnetic, strong, weak, and gravity. Each type of force has its own associated force carrier particles that are responsible for transmitting the force in particle interactions.

The first force on the list, and probably the most familiar, is electromagnetic. Electromagnetic force binds electrons to the nucleus to form electrically neutral atoms. It also is responsible for binding atoms into molecules. The carrier particle for electromagnetic force is the photon. Photons of different energies span the entire electromagnetic spectrum, from the highest frequency gamma rays to the lowest frequency radio waves. As far as physicists have been able to determine, photons have no mass and they travel at the speed of light.

The strong force (also called the strong interaction) exerts its effects only over incredibly small distances between subatomic particles. But at those small distances, it is very powerful. The strong force holds the nucleus together. It is so strong that it can overcome the electromagnetic force that makes positively charged protons repel one another. The carrier particles of the strong force are gluons. Gluons have no mass and no electric charge.

The weak force (also called the weak interaction) also acts only over very small distances. As its name implies, it is much weaker than the strong force. Nevertheless, all quarks and all leptons feel its effect. The weak force is responsible for radioactivity and for the decay of more massive quarks and leptons to produce lighter quarks and leptons. This decay process explains why all the stable matter in the universe contains only electrons and the lightest two quark types (up and down)—these particles cannot decay any further.

The carrier particles of weak force are the W and Z bosons. Unlike photons and gluons, bosons do have mass. The W boson, for example, is about eighty times more massive than a proton.

Finally, we have the fourth fundamental force: gravity. Everyone knows that gravity is the force that keeps our feet planted on the ground and keeps Earth revolving around the Sun. And physicists know that this force also must play a role in atomic structure and the interactions of subatomic particles. But the Standard Model cannot satisfactorily explain gravity. The force carrier for gravity—the graviton—has not yet been found.

While the effect of gravity on large objects such as people and planets is significant, gravity's effect on small objects such as subatomic particles is insignificant. Compared to the other fundamental forces, gravity appears not to be very important in the atomic world. So the Standard Model works without having to take gravity into consideration.

That's it in a nutshell: six quarks, six leptons, and four force carriers. This collection of subatomic particles and force carriers represents the set of essential building blocks required to make all the matter in the universe.

Is the Standard Model the final answer to explaining the fundamental nature of matter? No, certainly not. It simply represents the best explanation that scientists have put together so far. Like other theories in science, it will undoubtedly be modified as new information about subatomic particles and their behavior comes to light.

In 2003 two teams of researchers reported catching a glimpse of a new particle, formed in a particle accelerator, made up of five quarks. Called the pentaquark, it appears to be made of two up quarks, two down quarks, and an antistrange quark (the antiparticle equivalent of a strange quark). The pentaquark decayed to form a meson and a neutron in just 10^{-20} seconds!

The Standard Model also predicts the existence of a mysterious particle that has so far eluded the many physicists who are searching for it. This particle, called

STRING THEORY AND LOOP QUANTUM GRAVITY

The fact that the Standard Model does not explain gravity has been a thorn in the side of theoretical physicists since the model was developed. And in science, such "thorns" tend to lead to new ideas. Superstring theory, or string theory, attempts to reconcile gravity with the quantum theory of matter.

In string theory, there are no particles. Instead, there are minute vibrating "strings" floating in space-time (a four-dimensional world in which space and time are one). Strings are about a millionth of a billionth of a billionth of a billionth of a centimeter long. We can't see them, but we can see their effects. For example, if a string vibrates in one way, we see an electron. If it vibrates some other way, we see a quark or a photon. Strings can also break apart and recombine in ways that give rise to interactions between particles.

String theory includes more than strings too. There are also sheetlike structures, or branes (from the word *membranes*). They exist in a space-time world that contains many more dimensions than the four we are familiar with (height, width, length, and time). String theory solves some of the problems quantum mechanics can't and—very importantly—predicts particles that carry gravity.

That said, string theory may not be the final answer either. Another recently developed theory, loop quantum gravity, takes a different approach. It holds that everything is built up from a network of relationships or connections. When part of the network is tied in a braid, it forms something like a particle. Mathematically, some of the different braids match known particles, down to their spin and charge. Some theoretical physicists think that loop quantum gravity is an improvement on string theory. Whether that's true, and whether either one will replace the Standard Model, remains to be seen.

the Higgs particle, or Higgs boson, supposedly drags on other particles, thereby endowing them with mass. So far, no convincing evidence of the Higgs particle has been found. But scientists are hoping that a new particle accelerator will help them find it. The Large Hadron Collider (LHC) outside Geneva, Switzerland, is scheduled to begin operating in 2007. It is a huge circular accelerator, nearly 17 miles (27 kilometers) in diameter. When completed, the LHC will be the world's most powerful particle accelerator.

If or when the Higgs particle is found, particle physicists won't be out of a job. Despite its success, the Standard Model still leaves many questions unanswered. For instance, are quarks and leptons really fundamental, or do they have smaller parts? Are there more types of particles and forces that only higher energy accelerators will reveal? And where does gravity fit in?

In the twenty-five centuries that have passed since the ancient Greeks first pondered the nature of matter, we have learned a great deal. But many questions remain, and the search for answers will likely continue far into the future.

Glossary

alpha particle: a fast-moving helium ion, consisting of two protons and two neutrons, emitted by certain types of radioactive substances

antimatter: matter composed of antiparticles

atom: a building block of matter; the smallest unit of an element that still retains its identity in chemical reactions

atomic number: the number of protons in the nucleus of an atom

atomic weight: the average mass of the atoms (including all isotopes) in a naturally occurring element

beta decay: a form of radioactive decay in which an atom emits an electron or a positron, along with an antineutrino or a neutrino

beta particle: another (older) name for an electron

cathode rays: streams of fast-moving electrons emitted by the negative electrode (cathode) in a cathode ray tube

cathode ray tube: a vacuum tube with electrodes at each end originally used to study electricity

electrolysis: the process of using an electric current to separate a compound into its component elements

electromagnetic radiation: waves of energy that move through space and have both electric and magnetic properties

electron: a negatively charged subatomic particle that is found outside the atomic nucleus; one of six types of leptons

elements: substances that cannot be physically or chemically broken down into simpler substances

force carriers: subatomic particles responsible for transmitting force in particle reactions

frequency: the number of times a wave of electromagnetic radiation repeats per unit of time

gamma ray: a powerful form of high-frequency, short-wavelength electromagnetic radiation

gravity: the force of attraction between objects with mass

hadron: a subatomic particle made up of quarks that is sensitive to the strong force

ions: atoms that have gained or lost one or more electrons and carry a charge

isotopes: atoms of the same elements that have different atomic weights because they have different numbers of neutrons

lepton: an elementary subatomic particle that cannot be broken down; sensitive to the weak force

mass: the amount of matter something contains

molecule: a combination of one or more atoms chemically bonded together

neutrino: a massless or nearly massless subatomic particle that is electrically neutral

neutron: a neutral particle in the nucleus of an atom, with roughly the same mass as a proton

nucleus: the small, massive, positively charged center of an atom formed from protons and neutrons (the hydrogen nucleus contains only a proton)

orbitals: zones within an atom in which an electron is likely to be found

photon: a particle of light energy

positron: a positively charged electron; the antiparticle of an electron

proton: a positively charged particle in the nucleus of atoms

quantum: a small unit of energy; quanta is the plural form of quantum

quark: a fundamental subatomic particle; one of the most basic building blocks of all matter

radioactivity: the spontaneous break down, or decay, of the nucleus of an atom

spectral lines: a distinctive patterns of colored lines (specific wavelengths of light) given off by elements as they are heated

strong force/interaction: the force holding the nucleus of an atom together; transmitted by gluons

subatomic particle: a particle smaller than an atom; one of its component parts

transmute: to change one substance into another

wavelength: the distance between repeating parts (crests and troughs) of a wave pattern

wave-particle duality: having both wave and particle characteristics

weak force/interaction: the fundamental force responsible for radioactive decay processes; transmitted by bosons

X-ray: a penetrating form of high-frequency, short-wavelength electromagnetic radiation; only gamma rays are more powerful

TIMELINE

ca. 400 B.C. Greek philosophers Leucippus and Democritus propose that all matter is formed from atoms, a philosophy known as atomism.

ca. 50 B.C. Greek poet Lucretius writes *On the Nature of Things*, a poem restating the concepts of atomism.

A.D. 1492 Christopher Columbus makes his first voyage of discovery to the Americas.

1661 Robert Boyle publishes *The Sceptical Chymist*, which proposes that the true elements are substances such as sulfur and mercury—not earth, air, fire, and water.

1785 Antoine-Laurent Lavoisier proposes the law of conservation of mass. It states that mass is neither lost nor gained during a chemical reaction.

1794 Joseph-Louis Proust proposes the law of definite proportions, which states that the ratio of elements in a given chemical compound is constant.

1800 William Nicholson and Anthony Carlisle discover electrolysis.

1803 John Dalton publishes the first atomic theory.

1855 Johann Heinrich Wilhelm Geissler invents Geissler tubes, later known as cathode ray tubes.

1864 James Clerk Maxwell develops four equations based on Michael Faraday's work on electromagnetism.

1870 Dmitry Ivanovich Mendeleyev proposes organizing all known elements in a periodic table of the elements.

1895 Wilhelm Conrad Roentgen discovers X-rays.

1897 J. J. Thomson discovers the electron.

1898	Antoine-Henri Becquerel discovers radioactivity.
	Pierre and Marie Curie announce their discovery of the radioactive elements polonium and radium.
1905	Albert Einstein publishes his special theory of relativity, including $E=mc^2$. In the same year, he explains the photoelectric effect.
1909	Ernest Rutherford discovers the proton and the atomic nucleus.
1913	Niels Bohr proposes a new atomic model of hydrogen with mathematically defined "permitted orbits" for the electron.
	Henry Moseley uses X-ray diffraction to measure the amount of positive charge in atomic nuclei. This discovery leads to reorganization of the periodic table of elements.
1914–1918	World War I is fought.
1923	Louis de Broglie proposes that all forms of matter exhibit both wave and particle characteristics.
1925	Wolfgang Pauli develops the Pauli exclusion principle. It states that no two electrons in the same atom can be in exactly the same place at the same time with the same spin.
	Erwin Schrodinger devises a mathematical equation to describe the behavior of electrons for all atoms.
1932	James Chadwick discovers the neutron.
	John Douglas Cockcroft and Ernest T. S. Walton, working with Ernest Rutherford, build the first proton accelerator.
	In a cloud chamber, Carl David Anderson records the passage of a positively charged electron, the positron.

1939–1945	World War II is fought. The war ends when the United States drops atomic bombs on the Japanese cities of Hiroshima and Nagasaki.
1947	The first strange particle, the kaon, is discovered.
	The pion is discovered.
1955	The antiproton is discovered at Lawrence Berkeley National Laboratory in Berkeley, California.
1956	The neutrino is discovered.
1964	The omega-minus particle is discovered.
	Murray Gell-Mann and George Zweig, working independently, propose that subatomic particles are all made from a small number of truly fundamental particles, which come to be known as quarks.
1977	The bottom quark is discovered.
1995	The elusive top quark is discovered at Fermilab.
1996	Scientists at the European Organization for Nuclear Research (CERN) and Fermilab successfully create an antiatom of antihydrogen.
2003	The pentaquark—made up of two up quarks, two down quarks, and an antistrange quark—is observed by two research teams.
2007	The Large Hadron Collider is scheduled to open late in the year. It is a circular tunnel 17 miles (27 kilometers) in circumference, straddling the border between France and Switzerland.

Biographies

Niels Bohr (1885–1962) Born and educated in Denmark, Niels Bohr headed to the Cavendish Laboratory at Cambridge in 1911 to work on atomic structure under J. J. Thomson. Bohr applied the new idea of quanta to Rutherford's atomic model and introduced the idea that electrons lost or gained energy when they jumped between set orbits around the nucleus. The Bohr model of the atom, although later improved upon, was a huge step forward in understanding atomic structure. Bohr received a Nobel Prize for his work on the model in 1922. In the early 1940s, both Bohr and his son Aage joined the group of scientists working at Los Alamos National Laboratory in New Mexico. The scientists designed the two atomic bombs that were ultimately dropped on Japan in 1945. Troubled by his role in having created such a weapon, Bohr went on to support arms control after World War II. Bohr also helped establish the European Organization for Nuclear Research (CERN).

Marie Curie (1867–1934) Born in Warsaw, Poland, Marie Sklodowska Curie received some of her first scientific instruction from her father, a secondary school teacher. In 1891 she left Poland for Paris to study chemistry, mathematics, and physics at the Sorbonne. After completing her degrees, she became an instructor there. In 1895 she married Pierre Curie, who was also a physics instructor at the Sorbonne. Together the Curies carried out landmark experiments on radioactive elements. They isolated two new radioactive elements, polonium (named for Poland) and radium (named for its powerful radioactivity). Together with her husband and Antoine-Henri Becquerel, Marie Curie was awarded the Nobel Prize in Physics in 1903 for work on radiation. She was the first woman recipient of the prize. In 1911 she won the Nobel Prize in Chemistry, making her one of only two people who have been awarded the prize in two different fields. The Curies' daughter Irène continued their work with her husband, Frédéric Joliot-Curie. That couple won the Nobel Prize in Chemistry in 1935.

JOHN DALTON (1766–1844) The son of a Quaker weaver, John Dalton was born in Cumberland, England. He became a teacher at the age of twelve, focusing on the fields of mathematics and natural philosophy. He developed a fascination with meteorology and kept a daily diary of weather observations. Around 1800 Dalton turned his attention to chemistry. His early studies on gases led to development of the law of partial pressures (known as Dalton's law). Dalton devised a system of chemical symbols for the elements and determined the relative weights of some types of atoms. His conclusion that elements combine in simple numerical ratios by weight to form compounds led to the development of the law of definite proportions. His scientific masterpiece was his atomic theory. It held that elements are made of tiny, indestructible particles called atoms and that atoms of any given element are identical.

DEMOCRITUS (460–370 B.C.) A student of Leucippus, Democritus was an ancient Greek scholar, mathematician, and philosopher. He built on his teacher's idea that all matter is made up of particles called atoms. He believed atoms to be indestructible and indivisible and that the entire universe consisted entirely of empty space and atoms, which differed from one another only in size, shape, and arrangement. Born into a wealthy family in the Greek city of Abdera, Democritus was revered for his knowledge of natural phenomena and famous for being able to predict the weather. Democritus is thought to have authored around seventy books. None of his writings survived into later centuries, but his ideas were transmitted through the philosopher Epicurus and others.

MICHAEL FARADAY (1791–1867) One of the greatest experimental scientists of the nineteenth century, Michael Faraday was born into a poor and strongly religious family near London, England. At fourteen he became an apprentice bookbinder. He did more than just bind the books—he also read them, developing an intense interest in scientific topics, especially electricity and chemistry. A stroke of good luck enabled Faraday to attend a series of chemistry lectures by Humphry Davy. In 1812 Davy hired Faraday as an assistant, a relationship that continued for many years. Faraday went on to make a series of discoveries about chemistry, electricity, and magnetism that revolutionized physics. His ideas of electric and magnetic lines of force strongly influenced the subsequent understanding of atomic structure.

MURRAY GELL-MANN (1929–) Born in New York City, Murray Gell-Mann enrolled as a student at Yale University at the age of fifteen (he had already taught himself calculus). He went on to do graduate work at the Massachusetts Institute of Technology and later joined the staff at the California Institute of Technology as a professor of theoretical physics. Gell-Mann is probably best known for bringing order to the particle zoo by proposing the existence of subatomic particles called quarks. He also introduced a new fundamental property of matter—strangeness—to go along with more familiar ones such as electric charge. His work on subatomic particles won him the Nobel Prize in Physics in 1969. Gell-Mann went on to help develop the idea of color force and, later, string theory. Known for a sharp mind and remarkable expertise in subjects that few people grasp easily, Gell-Mann has been nicknamed the Man with Five Brains.

ERNEST RUTHERFORD (1871–1937) Born just outside Nelson, New Zealand, to British parents, Rutherford received his first science book when he was ten years old. He was inspired to make a cannon out of a hat peg, a marble, and blasting powder. It worked beautifully and, luckily, didn't hurt anyone. Rutherford grew into an outgoing, brawny, and brilliant young man who won a scholarship to attend the University of Canterbury in Christchurch, New Zealand. He received a degree in mathematics and physics. In 1894 Rutherford applied for a rare scholarship to study anywhere in the British Empire. Another applicant won the scholarship but wasn't able to go. Rutherford was the alternate. He chose to work with J. J. Thomson at Cambridge University's Cavendish Laboratory. There he studied ions in gases, the photoelectric effect, X-rays and, later, alpha and beta rays. Rutherford left Cambridge for McGill University in Montreal, Canada, where he continued his research for ten years, particularly his studies on alpha rays. After returning to Great Britain in 1907, Rutherford set up a center to study radiation. With the help of graduate students, he discovered the atomic nucleus and proposed a new atomic model in which electrons orbited the nucleus like planets around the Sun. Rutherford was awarded the Nobel Prize in Chemistry in 1908.

J. J. THOMSON (1856–1940) Sir Joseph John (J. J.) Thomson was born in Manchester, England. He is best known as the discoverer of the electron, an achievement that revolutionized the understanding of atomic structure. The son of a British bookseller, Thomson entered what is now Victoria University of Manchester at fourteen. He then received a scholarship to attend Trinity College at the University of Cambridge. Thomson stayed at Trinity for the rest of his life. In 1884 he was chosen to head Cambridge's Cavendish Laboratory, where some of the most cutting-edge research on atomic structure was conducted during the late 1800s and first several decades of the 1900s. Despite Thomson's experience in carrying out experiments, some of his graduate students found the scientist to be surprisingly clumsy with his hands and encouraged him not to touch their laboratory instruments! Thomson was awarded the Nobel Prize in Physics in 1906.

Source Notes

24   J. J. Thomson, *Recollections and Reflections* (London: G. Bell and Sons, 1936), 341.

Selected Bibliography

Asimov, Isaac. *Atom: Journey Across the Subatomic Cosmos.* New York: Dutton, 1991.

———. *The History of Physics.* New York: Walker, 1966.

Close, Frank, Michael Marten, and Christine Sutton. *The Particle Explosion.* New York: Oxford University Press, 1987.

Newton, David E. *Recent Advances and Issues in Physics.* Phoenix: Oryx Press, 2000.

Schumm, Bruce. *Deep Down Things: The Breathtaking Beauty of Particle Physics.* Baltimore: Johns Hopkins University Press, 2004.

Segre, Emilio. *From X-Rays to Quarks: Modern Physicists and Their Discoveries.* San Francisco: W. H. Freeman, 1980.

Von Baeyer, Hans Christian. *Taming the Atom: The Emergence of the Visible Microworld.* New York: Random House, 1992.

Watson, Andrew. *The Quantum Quark.* New York: Cambridge University Press, 2004.

Weinberg, Steven. *The Discovery of Subatomic Particles.* New York: W. H. Freeman, 1983.

Further Reading

Emsley, John. *Nature's Building Blocks: An A-Z Guide to the Elements.* New York: Oxford University Press, 2001.

Fleisher, Paul. *Relativity and Quantum Mechanics: Principles of Modern Physics.* Minneapolis: Twenty-First Century Books, 2002.

Fox, Karen. *The Chain Reaction: Pioneers of Nuclear Science.* New York: Franklin Watts, 1998.

McPherson, Stephanie Sammartino. *Stephen Hawking.* Minneapolis: Twenty-First Century Books, 2007.

Miller, Ron. *The Elements: What You Really Want to Know.* Minneapolis: Twenty-First Century Books, 2006.

Stwertka, Albert. *The World of Atoms and Quarks.* New York: Twenty-First Century Books, 1995.

Websites

American Institute of Physics
http://www.aip.org/history/
Under the heading "Exhibit Hall," you'll find information about Albert Einstein, Marie Curie, the discovery of the electron, and more.

The Nobel Prize in Physics
http://nobelprize.org/physics/educational
This site includes information on particle accelerators, matter, X-rays, and many other big ideas that led to Nobel Prizes.

The Official String Theory Website
http://www.superstringtheory.com/index.html
Learn all about string theory. Many sections of this site let you choose a basic explanation or an advanced explanation.

The Particle Adventure
http://particleadventure.org/particleadventure/index.html
View an animated slideshow that thoroughly explains the Standard Model with many specific examples. This website also contains a comprehensive glossary.

A Science Odyssey: You Try It: Atom Builder
http://www.pbs.org/wgbh/aso/tryit/atom/#
Build a carbon atom from quarks, and read about important discoveries that contributed to the creation of the Standard Model.

INDEX

Gell-Mann, Murray, 54–55, 57, 74
gravity, 62–63, 64, 65

hadrons: baryons, 60; mesons, 60, 63
Higgs particle (Higgs boson), 65

ions, 18, 26, 36

kaon, 51, 60, 71

laws: Boyle's law, 12; law of conservation of mass, 12–13, 69; law of definite proportions, 13, 69; Pauli exclusion principle, 42, 44, 61, 70
leptons, 58, 60, 61, 62, 63, 65

matter, 4–6, 7, 8, 24, 33, 54, 58–65. *See also* antimatter; atoms
muon, 50, 51, 52, 60

neutrino, 52, 60, 71
neutrons, 46, 49, 51, 60, 63, 70; charge, 56. *See also* hadrons

particle accelerator, 47, 49, 50, 53, 55, 56, 57, 63, 65, 70, 71
Pauli, Wolfgang, 42, 52, 70
periodic table. *See* elements
photons, 33, 38, 61–62, 64
pion, 51, 52, 60, 71
positrons, 49, 51, 53, 54, 71
protons, 51, 53, 55, 60, 61, 62, 70; charge, 56; cloud chamber, 48, 49, 51, 71; gold foil experiment, 34–37; as nuclear particles, 40, 45, 46; proton accelerator, 47, 49, 53, 70; quark experiments, 56–57. *See*

also antiproton; hadrons
quantum numbers, 38, 42
quarks, 54–57, 58–60, 61, 63, 64, 65, 71; color, 59, 61; pentaquarks, 63, 71. *See also* hadrons

radioactivity, 30–31, 70
rays, radioactive: alpha, 30, 31, 34; beta, 30, 31; gamma, 30–31, 45, 51, 61
Roentgen, Wilhelm Conrad, 27–29, 69
Rutherford, Ernest, 31, 34, 35, 36, 40, 43, 45, 47, 49, 70, 74–75

Schrodinger, Erwin, 41–42, 43, 70
spectral lines, 37, 38, 40
Standard Model of Fundamental Particles and Interactions, 58, 61, 63, 64, 65, 71

theories: loop quantum gravity, 65; quantum theory, 32, 33, 37, 38, 42, 44, 55, 64; Standard Model of Fundamental Particles and Interactions, 58, 61, 63, 64, 65, 71; strangeness, 55; string theory, 64; theory of light quanta, 33, 37, 50; wave-particle duality, 40–41, 70
Thomson, Joseph John (J. J.), 21–22, 24, 25, 26, 70, 75

ultraviolet (UV) light, 20, 25, 29

wave-particle duality, 40–41, 70

X-rays, 28–29, 30, 69; X-ray diffraction, 39

Photo Acknowledgments

The images in this book are used with permission of:

© Getty Images, p. 5; © Laura Westlund/Independent Picture Service, pp. 11, 20, 22, 26, 35, 43, 56, 59; Courtesy of the National Library of Medicine, p. 28; © LBNL/Photo Researchers, Inc., p. 48.

Cover: © Tim Parlin/Independent Picture Service.